DON'T BURST THE EGG

Author LaReina Lewis

First published in UNITED KINGDOM, NOVEMBER 2020
Copyright ©LAREINA LEWIS

All rights reserved. No part of this publication maybe reproduced, stored in a retrieval system or transmitted in any form or by any means without the prior permission in writing of the publisher, nor be circulated in writing of any publisher, nor be otherwise circulated in any form of binding or cover other than that in which it is published without a similar condition including this condition, being imposed on the subsequent purchaser.

This book has been produced for the Amazon Kindle
and is distributed by Amazon Direct Publishing

Cover created by LaReina Lewis

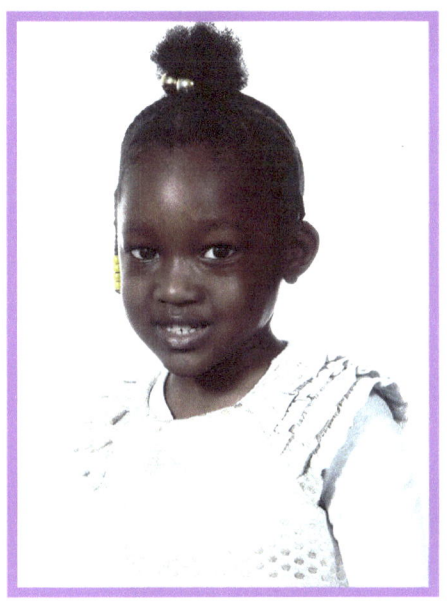

LaReina Lewis, age 5, lives with her mother and father in London. LaReina loves nature, animals, swimming, crafting , drawing, singing, science experiments and reding with her teacher.

After conducting the science experiment of removing the carbon from an egg and reading a selection of school library books, LaReina was inspired to create this book.

We hope you enjoy reading this fun story. Give the experiment and try and remember, DON'T BURST THE EGG!

Otha Lorelai

They all went to the sink.

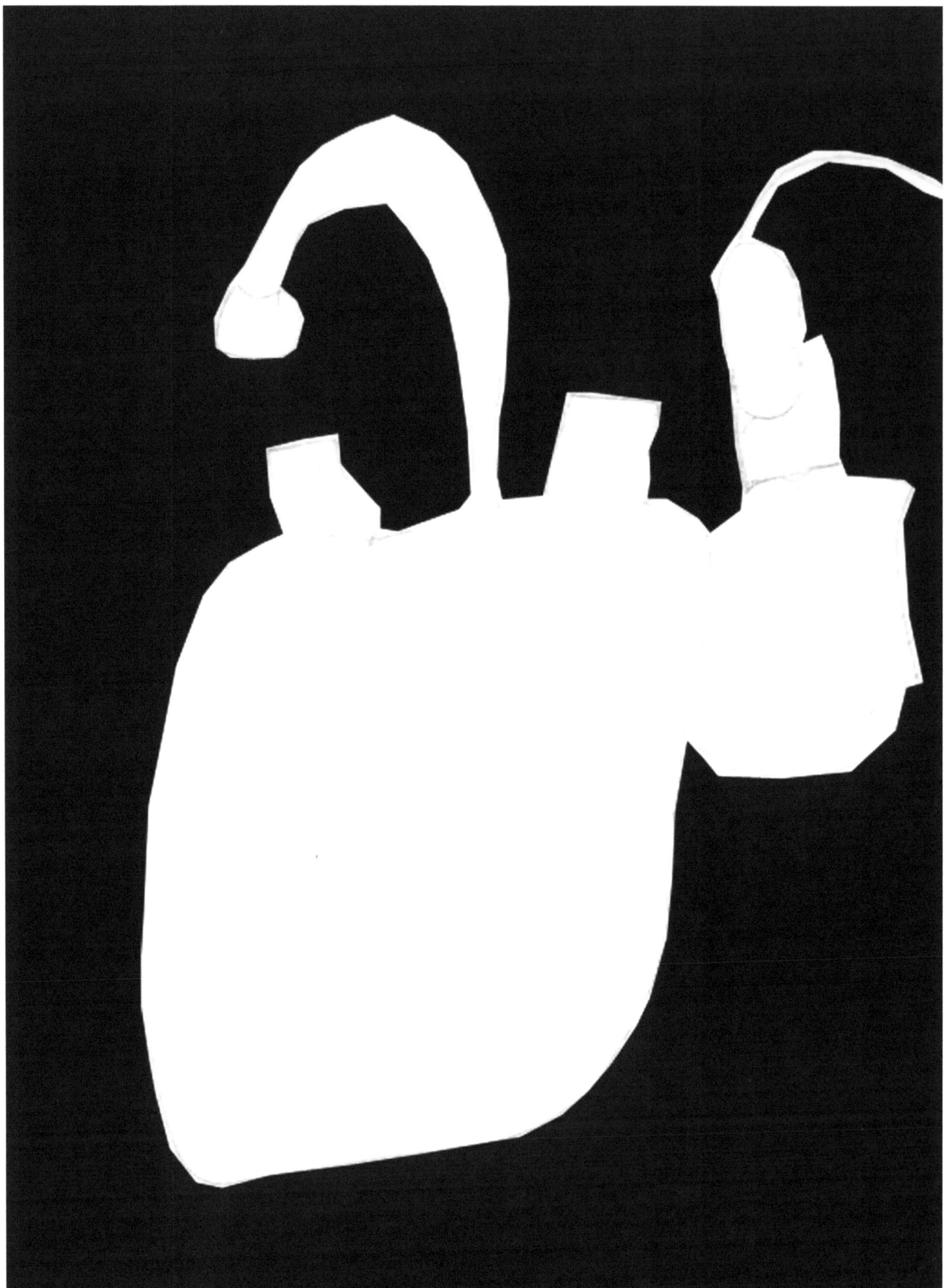

ZZZZZ

They snored.

In the morning...

They all went to check on the egg.

When Sis started to take

the egg out of the cup

of vinegar,

it was

all

SQUISHY!

"Be careful!"
Dotspot said, as Sis gave Pigeon the squishy egg.

"Oh Yessss, Yesssss" Pigeon said.

THE END

THE END
THE
STE

by Lahaina

DON'T BURST THE EGG

This book tells the story of Pigeon, Sis the owl and Dotspot the penguin, who decide to conduct a science experiment. Find out what happens when the carbon is removed from an eggshell.

Author LaReina Lewis

www.ingramcontent.com/pod-product-compliance
Lightning Source LLC
Chambersburg PA
CBHW040057250526
45473CB00043B/1853